What Is the Impact of
OCEAN POLLUTION?

Other titles in the *Environmental Impact* series include:

What Is the Impact of Climate Change?
What Is the Impact of Declining Biodiversity?
What Is the Impact of Excessive Waste and Garbage?

Environmental
IMPACT

What Is the Impact of
OCEAN POLLUTION?

Craig E. Blohm

ReferencePoint
Press

San Diego, CA

© 2021 ReferencePoint Press, Inc.
Printed in the United States

For more information, contact:
ReferencePoint Press, Inc.
PO Box 27779
San Diego, CA 92198
www.ReferencePointPress.com

LIBRARY OF CONGRESS CATALOGING-IN-PUBLICATION DATA

Names: Blohm, Craig E., 1948- author.
Title: What is the impact of ocean pollution? / by Craig E. Blohm.
Description: San Diego, CA : ReferencePoint Press, Inc. 2020. | Series:
  Environmental impact | Includes bibliographical references and index.
Identifiers: LCCN 2020012399 (print) | LCCN 2020012400 (ebook) | ISBN
  9781682828656 (library binding) | ISBN 9781682828663 (ebook)
Subjects: LCSH: Marine pollution--Juvenile literature. | Marine
  ecology--Juvenile literature.
Classification: LCC GC1090 .B56 2020 (print) | LCC GC1090 (ebook) | DDC
  393.739/409162--dc23
LC record available at https://lccn.loc.gov/2020012399
LC ebook record available at https://lccn.loc.gov/2020012400

Contents

The Endangered Oceans

The tiny submersible slowly made its way through the unrelenting darkness of the Mariana Trench, the deepest place on earth. It was April 2019, and the submersible, named DSV *Limiting Factor*, was piloted by retired naval officer and marine explorer Victor Vescovo. Located in the western Pacific Ocean, the 1,580-mile (2,543 km) trench had been visited by only a handful of explorers. Vescovo planned his dive to be the deepest in history and was eager to explore the Challenger Deep, the lowest spot in the trench.

After a nearly four-hour descent, the *Limiting Factor* reached its goal: 35,853 feet (10,928 m) below the surface, the lowest point a human being has ever gone. In the beams of the submersible's lights, Vescovo observed several new species of sea creatures never before seen by humans. He also discovered something else that all his research had not prepared him to find: trash. "It could have been plastic," says Vescovo, "and it could have been a plastic bag. But we definitely did see the lettering of some kind on it, so it was refuse."[1] Trash and other forms of ocean pollution are having a detrimental impact on all parts of the ocean environment and, if the threat is not taken seriously and confronted promptly, will have disastrous consequences for all inhabitants of the planet.

Polluting the Ocean Environment

For marine biologist and ocean explorer Sylvia Earle, the ocean has been a lifelong obsession. "As a child, I was aware of the widely-held attitude that the ocean is so big, so resilient that we could use the sea as the ultimate place to dispose of anything we did not want, from garbage and nuclear wastes to sludge from sewage to entire ships that had reached the end of their useful life."[2] Earle believes this attitude threatens the planet's most precious natural resource.

As Earle's statement indicates, ocean pollution takes many forms. In fact, most pollution originates on dry land, not in the oceans. Fertilizers, pesticides, and other agricultural chemicals spread on farmland are washed by rainstorms into streams and rivers, ultimately reaching the ocean. Millions of oil drops deposited by cars and trucks on parking lots and city streets every day can be washed into sewers and eventually reach the oceans as well. Air pollution emitted by factories and vehicles is yet another source of pollution since airborne emissions fall to the oceans, drawn by rainfall or gravity. Household and industrial chemicals (containing dangerous elements such as zinc and mercury), medical waste, and radioactive waste are often disposed of without considering that they, too, can make their way into waterways or that they have grave consequences on the marine ecosystem.

Visible Debris, Visible Damage

The most widespread source of ocean pollution, however, is plastic in its many forms. Plastic bags, drinking straws, nonreusable water bottles, and industrial and consumer goods are just a few examples of plastic pollution that have been found in every ocean on earth. It can take hundreds of years for plastics to degrade, and most never disappear completely. They break down into tiny

Workers use both chemical and mechanical processes to clean up an oil spill.

pieces called microplastics, which have been found in food and drinking water. About 8.8 million tons (8 million metric tons) of plastic is dumped into the oceans each year, endangering the lives of fish, marine mammals, and seabirds.

Seagoing vessels such as cruise liners and cargo ships, as well as offshore oil-drilling rigs also contribute to ocean pollution. Some ship crews dump garbage and waste overboard despite laws prohibiting such polluting; tanker ships and petroleum drilling rigs can leak oil into the oceans, creating a mess that is difficult to clean up. Despite having modern navigation equipment, ships sometimes collide with other vessels or run aground on rocky shores, releasing huge quantities of oil into the oceans. In 2018, 127,868 tons (116,000 metric tons) of oil were spilled worldwide due to tanker ship accidents.

Of all the oceans' creatures, coral reefs may be the first to suffer irreparable damage due to ocean pollution. Their loss has become a rallying point for those concerned about ocean pollution because the reefs are so important to the planetary environment and human well-being. In the past thirty years, more than one-quarter of the world's coral reefs have been destroyed due to pollution and other factors such as climate change. According to a National Geographic Society report, most reefs could be gone by the year 2100. This would affect more than 500 million people who depend on the reefs for income, food, and protection from coastal erosion.

Solving the Pollution Problem

The enormity of ocean pollution may be daunting, but efforts are already being made to address the problem. New technologies remove plastic pollution from the oceans, and oil spills are cleaned up by chemical and mechanical processes. The National Oceanic and Atmospheric Administration (NOAA) uses high-resolution satellite imagery to monitor the health of coral reefs and identify potential threats to these fragile ecosystems. Conservation organizations have established grassroots campaigns to remove garbage from shorelines, and they have encouraged the recycling and repurposing of plastic and other single-use items before they become a danger to the environment. Government and industry are creating ways to help keep the oceans from becoming further polluted. For example, by 2021 all single-use plastics will be banned in Canada. The Indonesian government has pledged up to $1 billion a year to clean up ocean pollution. Dell, Coca-Cola, Nestlé, Unilever, and other corporations are beginning to increase the use of recyclable packaging. And operators of tanker ships, oil rigs, and other petroleum facilities must demonstrate that they have a plan and the resources to clean up spills.

The fight against ocean pollution has intensified, but more work needs to be done and awareness raised of the dangers that face the earth's vital seas. "We need to respect the oceans," Earle warns, "and take care of them as if our lives depended on it. Because they do."[3]

Chapter One

Plastic Pollution

In the famous 1870 science fiction novel *Twenty Thousand Leagues Under the Sea*, a captain named Nemo explores the vast reaches of the underwater world in a futuristic submarine. Today there is a place in the southern Pacific Ocean that shares its nickname with the fictional captain. Point Nemo, officially known as the Oceanic Pole of Inaccessibility, is more than 1,000 miles (1,610 km) from the nearest land, making it the most remote place in the world's oceans. *Nemo*, which is Latin for "no one," is a fitting name for an area that very few people have ever visited.

During an around-the-world ocean race in 2018, sailors on the yacht *Turn the Tide on Plastic* took water samples as their boat sailed through Point Nemo. After sending the samples to Germany for analysis, they learned that there were twenty-six tiny pieces of plastic in every 35.3 cubic feet (1 cu. m) of seawater. "We've found microplastics, sadly, in nearly all of our samples," says Sören Gutekunst, science consultant for the ocean race, who analyzed the samples. "This shows how pervasive and vast the problem is already."[4]

A Far-Reaching Problem

Finding plastic in the oceans' most remote location is a troubling indication of the extent of plastic pollution in the seas. Numbers confirm the enormity of the problem. According to environmental advocacy organization Ocean Conservancy, every year 8.8 million tons (8 million metric tons) of plastic makes its way into the world's oceans; that equals the weight of 86 nuclear aircraft carriers or 4.4 million cars. There were approximately 165 million tons (150 million metric tons) of plastic waste (including tiny fragments called microplastics) in the oceans by 2020. A study by the World Wildlife Fund estimates that plastic pollution could show a twofold increase in less than ten years. "If business continues as usual," the report states, "by 2030 the plastic system is expected to double the amount [of] plastic pollution on the planet, with oceans the most visibly affected. Although existing initiatives to combat plastic pollution are in place in many regions, they are not enough."[5] Another study by the British charity the Ellen MacArthur Foundation predicts that by 2050 there could be more plastic, by weight, than fish in the oceans.

Plastic has become such an essential part of modern life that it is no surprise its use is increasing. The first true plastic, called Bakelite after its creator, Leo Baekeland, was invented in 1907. By the 1950s, when large-scale plastic production began, millions of products were being made from this versatile material. Today one of the most common types

IMPACT FACTS

The Great Pacific Garbage Patch weighs 87,000 tons (78,925 metric tons)— more than forty-three thousand automobiles.

—Public Broadcasting Service

of plastic products, and one of the most harmful to the environment, is the plastic bags available at almost any grocery store. "The average American family takes home almost 1,500 plastic shopping bags a year," says Eric A. Goldstein of the nonprofit Natural Resources Defense Council, "clogging our cabinets, kitchen drawers and landfills. They're hanging from trees, and littering our beaches."[6] While statistics vary widely, as many as 5 trillion plastic bags are used worldwide every year. The average amount of time a plastic bag is used is twelve minutes, while carrying purchases from store to home. After that, the bags are usually discarded; only 1 percent of all plastic bags are recycled.

Since 1988 plastic items have included a recycling symbol and a number that indicates the type of plastic in that item. The number aids in proper sorting for recycling, and some recycling services will not accept all numbered plastics because they can be difficult or dangerous to recycle. Still, it is ultimately up to industry and individuals to commit to recycling their used plastics. And therein lies the problem. A 2017 study published in the journal *Science Advances* discovered that, of all the plastic ever created, only 9 percent has been recycled. According to the study, about 12 percent is incinerated, and the rest, 79 percent, ends up in landfills or is simply left to litter the global environment.

Plastic Trash Reaching the Oceans

Although most of the 79 percent of plastic not recycled or burned is disposed of on land, it does not always stay where it is put. During the transporting of plastic refuse by truck to a landfill, some of the plastic can be dropped or blown away before it gets to its destination. Even after the load reaches the landfill, the wind can carry lightweight plastic trash far from the dumping site. Careless disposal, rather than recycling, of used or unwanted plastic by businesses and households also allows plastic refuse to enter the environment.

Along with plastic bags, billions of other single-use plastic items are discarded every year. Organizations such as NOAA and Ocean Conservancy have compiled lists of the most abundant plastic ocean pollutants. These include food wrappers, beverage bottles, bottle caps, takeout containers, drink lids, straws, single-use utensils, six-pack rings, and cigarette butts, which contain plastic fibers in their filters. Eventually, a lot of this trash finds its way to the world's oceans, creating an ecological disaster that threatens to devastate the marine environment. One of the routes that carries plastic pollution to the oceans is the rivers that ultimately empty into the sea.

Rainstorms can drive plastic litter that has not been properly disposed of or recycled into nearby rivers. As hydrogeologist Christian Schmidt notes, "The more waste . . . that is not disposed of properly, the more plastic ultimately ends up in the river

Plastic Breaks Down More Slowly than Most Other Ocean Debris

Of the many types of manmade objects that pollute the oceans, plastic is the worst offender. In some cases it takes hundreds of years to biodegrade. Fishing line, plastic bottles, and disposable diapers are among the most long-lasting of all ocean debris.

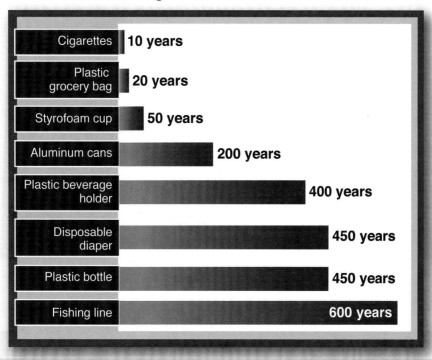

Estimated Number of Years for Selected Items to Biodegrade in a Marine Environment

Cigarettes	10 years
Plastic grocery bag	20 years
Styrofoam cup	50 years
Aluminum cans	200 years
Plastic beverage holder	400 years
Disposable diaper	450 years
Plastic bottle	450 years
Fishing line	600 years

Source: Kate Whiting, "This Is How Long Everyday Plastic Items Last in the Ocean," World Economic Forum, November 2, 2018. www.weforum.org.

and takes this route to the sea. . . . The quantity of plastic per cubic meter of water [is] significantly higher in large rivers than small ones."[7] In a 2017 report coauthored by Schmidt, ten rivers were identified as accounting for 88 to 95 percent of all plastic that makes its way from rivers to the sea. These rivers are all located in Asia and Africa, continents that have immense coastal populations and often lack proper facilities for waste management.

Rubber Ducks at Sea

During a storm in 1992, a shipping container was washed overboard from a cargo ship in the northern Pacific Ocean. Breaking open, the container disgorged its load: 28,800 plastic duck, frog, and turtle bath toys. As the toys floated away, the yellow ducks would become a symbol of the extent of plastic pollution in the oceans.

After about ten months, the ducks (popularly referred to as rubber ducks despite being plastic) began showing up on beaches around the world. Propelled by the ocean currents that create the ocean gyres, which are the great ocean swirls in various parts of the world, ducks arrived in Alaska, Australia, South America, the eastern United States, and Europe. Some were even found trapped in Arctic polar ice. In all, the celebrated ducks, which have garnered the nickname Friendly Floatees, have traveled tens of thousands of miles around the world.

Several thousand rubber ducks stuck in the North Pacific Gyre have helped researchers learn more about the Great Pacific Garbage Patch, a swirl of plastic debris said to be larger than Texas. Oceanographers have gained valuable information about worldwide ocean currents as the little yellow travelers wash up on remote shores. And the fact that the ducks continue to appear years after entering the ocean in 1992 attests to the long life span of plastic and its continued presence in the ocean.

Plastics Discarded by Ocean Vessels

Seagoing vessels are another source of plastic ocean pollution. A small, uninhabited island in the South Atlantic Ocean has shown evidence of plastic trash washed ashore after being discarded from merchant ships. Named Inaccessible Island by early explorers, the island, an extinct volcano, lies near shipping routes sailed by freighters from Asia to South America. A 2018 study of trash on Inaccessible's coastline revealed crushed plastic bottles with their caps still in place—a standard practice for saving space on ships. In addition, date stamps on the bottles indicated they were manufactured within the previous two years, meaning they came from a source nearby, rather than being brought from a long distance by ocean currents. "It's inescapable that it's from ships, and it's not coming from land," says Peter Ryan, lead author of the study. "A certain sector of the merchant fleet seems to be doing that, and it seems to be largely an Asian one."[8]

Shipping vessels are not the only culprits of ocean pollution. Fishing boats also contribute to ocean litter by dumping their used fishing nets and other gear overboard. Once made of biodegradable materials such as tree fibers and cotton, fishing nets are now almost exclusively made of plastic. Every year, some 705,500 tons (640,018 metric tons) of nets and other fishing gear are lost due to rough weather or are discarded, often to cover up illegal fishing activities. This "ghost gear" can linger in the water for hundreds of years, presenting a hazard to marine life. Plastic nets can extend for miles, and even after they are abandoned, they can continue to catch fish, a process known as ghost fishing.

The Great Pacific Garbage Patch

Ghost fishing nets make up almost half of a floating mass of ocean litter known as the Great Pacific Garbage Patch (GPGP). In 1997 oceanographer Charles Moore discovered the GPGP in the North Pacific Ocean as he sailed his yacht through the area after competing in an ocean race. He colorfully describes the phenomenon he saw: "It was and is a thin plastic soup, a soup lightly seasoned with plastic flakes, bulked out here and there with 'dumplings': buoys, net clumps, floats, crates, and other 'macro debris.' . . . This immense section of the northeastern Pacific Ocean, about halfway between Hawaii and The West Coast, was strewn throughout with buoyant plastic scraps."[9]

IMPACT FACTS

Seventy-three percent of all beach trash around the world is plastic.

—National Geographic Society

Moore estimated that the "soup" was approximately the size of Texas (although twenty years later, it had reportedly doubled in size). Media reports soon brought the GPGP to the attention of the general public, and it became a vivid example of the way humans are impacting the oceans.

The GPGP is only one of five such accumulations of trash located around the world. Theses patches are created by swirling ocean currents called gyres. "Gyres are large systems of circulating ocean currents, kind of like slow-moving whirlpools," says Megan Forbes, a marine biologist and communications consultant for NOAA. "There are five gyres to be exact—the North Atlantic Gyre, the South Atlantic Gyre, the North Pacific Gyre, the South Pacific Gyre, and the Indian Ocean Gyre—that have a significant impact on the ocean."[10] The GPGP is the largest of the five patches, containing an estimated 1.8 trillion pieces of plastic. Although Moore's description of the GPGP conjures up images of a massive floating garbage dump, in reality it is very difficult to observe the litter there. While some of the plastic litter floats on the waves, much more of it sinks below the surface, creating a column of debris that extends to the ocean floor. TVs, tires, nets, and other heavy items can be found at the sea bottom.

Microplastics in the Ocean

Along with all the larger pieces of plastic debris, the GPGP and other gyres also contain millions of smaller bits of plastic that, despite their size, can be extremely harmful to the ocean environment. Microplastics are small plastic pieces measuring about 0.2 inches (5 mm) or less in length. These particles do not start out as tiny bits but are the result of the physical degradation of larger plastic debris. Plastic is not biodegradable like paper, cloth, and other organic items. But sunlight and the action of waves cause plastic to break down into smaller and smaller pieces, eventually turning them into microplastics. Bottles, plastic tableware, consumer packaging, and even contact lenses that are disposed of in the sink or toilet can become microplastics.

Along with the microplastics caused by natural decomposition, there are other plastics that are manufactured to be extremely small. Microbeads are tiny plastic beads that are used as abrasives in consumer products such as face wash, exfoliants, and toothpaste. Nurdles are small plastic pellets used as

Plastic waste is not biodegradable, but sunlight and the action of waves causes items like bottles and consumer packaging to break into tiny pieces called microplastics.

raw material by manufacturers to make a multitude of products, from plastic bottles and caps to car dashboards and computer keyboards. As small as microplastics are, their sheer numbers create an ecological nightmare. While determining exact numbers is not possible, a study conducted in 2014 estimated that there were from 15 trillion to 51 trillion pieces of microplastics in the oceans, collectively weighing as much as 236,000 tons (214,096 metric tons).

Wildlife Ingesting Plastic Waste

With thousands of tons of plastics—ranging from minuscule microbeads to mile-long fishing nets—traveling the oceans, this pollution is a major threat to the marine ecosystem. Its most devastating effects are on the wildlife that inhabit the world's oceans.

In November 2018 a dead whale washed up on the shores of Indonesia. Researchers found over one thousand pieces of plastic in the stomach of the deceased cetacean. According to a *New York Times* article, the plastic trash in the whale's stomach included

The Voyage of the *Plastiki*

In 2006 thirty-one-year-old environmentalist and adventurer David de Rothschild had a crazy idea. He wanted to raise awareness of plastic pollution in the ocean by sailing across the Pacific Ocean, passing by the GPGP and filming a documentary about the journey. The crazy part? His boat would be made of 12,500 recycled plastic bottles as a symbol of the huge amount of plastic in the ocean and a reminder of the usefulness of recycling.

De Rothschild spent years designing and building his plastic boat, which he christened the *Plastiki*. On March 20, 2010, the 60-foot (18.3 m) catamaran, with de Rothschild and his crew of five aboard, set sail from San Francisco, California, beginning a journey that would ultimately take them to Sydney, Australia. During the four-month, 8,000-mile (12,875 km) voyage, the *Plastiki* encountered storms, battled high waves and suffocating heat, and sailed daily through lots of floating garbage.

By the time the *Plastiki* reached Sydney on July 26, 2010, the voyage was over, but the mission continued. "I think the hard work begins today," said de Rothschild upon docking. "It's about trying to explain this really dumb issue that we have—we've got this crazy problem with plastic in our ocean that is not abating. We've got this addiction to single-use plastics. I hope that the *Plastiki* will swing the needle and get people to recognize that we can solve this."

Quoted in Bradley Blackburn and Laura Marquez, "Trash Boat: 'Plastiki' Ship Made of Recycled Bottles Completes Voyage Across Pacific Ocean," ABC News, July 26, 2010. https://abcnews.go.com.

"115 cups, 25 bags, four bottles and two flip-flops."[11] In 2019 another whale that had washed up on a Scottish island was found to have "bundles of rope, plastic cups, bags and gloves,"[12] in its stomach, according to a *Washington Post* article. And it is not only large animals like whales that are being killed by plastic litter. There are more than eight hundred marine species vulnerable to injury or death by plastic. Sea turtles, fish, seals, seabirds, and dolphins can mistake microplastics for food and ingest harmful amounts of litter. This fills their stomachs and leaves no room for the nutrition they need to survive; they may suffer for months before eventually succumbing to starvation. About one hundred thousand turtles and other marine creatures die every year due to plastic litter.

The harm that plastic ocean pollution causes does not end with the sea creatures it affects. Toxic chemicals in the plastic eaten by fish and other aquatic animals leach into their bodies. When smaller fish are eaten by larger ones, the toxins transfer to the new host. Eventually, these fish can be caught and end up on a dinner plate; the toxins are then passed along to the humans enjoying the fish, unaware of the harmful chemicals in their meal. Some of these chemicals, such as bisphenol A and phthalates, have been linked to certain cancers, birth defects, respiratory problems, and other ailments. Humans also consume microplastics that are eaten by fish and other animals. A 2019 survey found that Americans consume as many as fifty-two thousand microplastic particles in their food each year.

Ghost Nets

Abandoned plastic fishing nets are also deadly for sea creatures. These ghost nets can travel thousands of miles via ocean currents, catching fish and other marine animals that have no way of escaping. Unable to swim normally while stuck in the nets, fish can become exhausted and face starvation and suffocation.

Abandoned plastic fishing nets are a deadly menace for fish and other sea creatures.

As these nets accumulate more and more fish, their weight can cause them to sink to the bottom of the ocean, where scavengers feed on the trapped creatures. Once the fish are gone, either by predators or natural decomposition, the nets rise, and the cycle of trapping begins again.

Larger sea creatures such as whales can also become ensnared in ghost nets. In trying to escape they may roll and twist, entangling themselves even further in the net. Swimming while entangled in a ghost net is stressful and exhausting for a whale and can make feeding difficult or impossible. Female whales trapped in nets are less likely to reproduce, further endangering the survival of the species. For one type of whale known as the right whale, entanglement in netting has become the leading cause of death, surpassing being struck by ships.

As a global problem, ocean plastic pollution affects everyone on the planet, making it a priority for drastic action. "What is concerning," notes Nicholas Mallos, senior director of the Trash Free Seas Program at Ocean Conservancy, "is that the likelihood of these types of interactions, and these types of horrific encounters between marine organisms and plastic debris, is only likely to continue unless some drastic measures are taken."[13]

Chapter Two

Oil Pollution

The East China Sea is a part of the Pacific Ocean surround-ed by China, Taiwan, and Japan and its Ryukyu Islands. Well-traveled by commercial shipping vessels, it is a dan-gerous area of the ocean where collisions between ships are not uncommon. On January 8, 2018, the Iranian oil tanker *Sanchi* collided with a Chinese cargo ship in the East China Sea. The tanker was carrying 150,000 tons (136,078 metric tons) of light crude oil, equivalent to about 1 million barrels of oil. The *Sanchi* drifted aimlessly until it exploded; four days later it sank to the bottom of the East China Sea. All thirty-two sailors on board the *Sanchi* perished.

The collision was a tragedy for the families of the *Sanchi*'s crew, but it was also an environmental disaster of huge pro-portions. The light oil carried by the tanker readily dissolves in water, making it difficult to clean up, and it is highly toxic to marine life. Many of these animals—including bluefin tuna, mackerel, squid, and crabs—use the East China Sea for their wintering ground. Along with the light oil carried by the *Sanchi*, the heavy bunker oil used as the ship's fuel can remain in the marine environment for long periods, washing up on coast-lines and endangering coastal birds and mammals.

The oil spill from the *Sanchi* in 2018 was not an isolated incident. According to NOAA, thousands of oil spills occur every year. Most of these are small, averaging perhaps only a ton of oil or less spilled. But in the last half century, at least forty-four major spills have occurred, averaging some 13,200 tons (11,975 metric tons) of oil. These numbers

The sinking of the Iranian tanker Sanchi *(shown here being doused with water by a Chinese fire and rescue boat)* created an environmental disaster after its cargo of light crude oil spilled into the East China Sea.

highlight the ongoing threat to the oceans posed by the accidental spilling of the petroleum products that keep modern society running. Cars, trucks, machinery, heating systems, and thousands of other products need oil to operate. But this reliance on oil brings with it the danger of polluting the environment.

Common Sources of Oil Pollution in the Ocean

One hazard that results from the world's dependence on oil products is the very fact that oil needs to be transported. Shipping oil from one country to another is a necessary part of the economic chain from oil producer to oil consumer. The world's oil tankers make around twenty-two thousand voyages annually, delivering 2.2 billion tons (2 billion metric tons) of oil to various destinations. With so much maritime activity, accidents are bound to happen. "It's a fact of life that there will be oil spills, as long as oil is moved from place to place,"[14] says Sylvia Earle, noted marine biologist and ocean explorer. In 1989 the tanker *Exxon Valdez* ran aground in Prince William Sound off the southern coast of Alaska. Its spill of 40,785 tons (37,000 metric tons) fouled some 1,300 miles (2,092 km) of the Alaskan coastline. The damage included the deaths of 250,000 seabirds, 250 bald eagles, 22 whales, and untold numbers of fish, which devastated the local fishing industry.

Ship accidents like those involving the *Sanchi* and the *Exxon Valdez* are not the only cause of oil spills in the ocean. In fact, NOAA reports that ship collisions and groundings account for only 12 percent of all the oil released into the oceans. Recreational boating—which includes fishing, water skiing, cruising, and the use of jet skis—is also a source of oil and fuel entering waterways. When oil is discharged from these watercraft, it is usually due to the negligence of the operator or a lack of education about the proper operation or maintenance of recreational marine vehicles.

The natural seepage of oil from undersea deposits, while not caused by human activity, is another major source of oil in the oceans. Because such seepages tend to be very old, however, bacteria in these areas have had time to evolve into microbes that consume much of this seeped oil. Thus, it is less hazardous to the environment than anthropogenic, or human-caused, oil spills.

IMPACT FACTS

A major offshore oil spill occurs about every twenty years.

—Ocean Conservancy

Drilling for Oil

As tanker ships ply the oceans with their liquid cargo, other crews are searching for oil located under the ocean floor. As with any human endeavor, accidents may occur during the process of drilling for these deep-sea oil deposits. Drilling rigs on the surface of the ocean extend drill bits to the rocky seafloor below, often at depths of 5,000 feet (1,524 m) or more. A vast amount of oil can be tapped by an oil rig, leading to the possibility of an environmental disaster if things go wrong. On April 20, 2010, just such a disaster occurred when the BP oil rig *Deepwater Horizon* exploded during a drilling operation in the Gulf of Mexico. For two days the rig burned, until it sank on April 22. The *Deepwater Horizon* had experienced a failure of a cement barrier at the undersea well, allowing oil and gas to escape and leading to a fiery explosion and the release of 682,000 tons (618,700 metric tons) of oil into the gulf. Eleven workers died in the explosion and fire on the *Deepwater Horizon*, and the disaster created an oil spill that continued to pollute the gulf for eighty-seven days.

Accidental spills pose major threats to the ocean's ecosystem. But the worst oil spill in history created by humans was not an accident but an act of war. In the 1991 Persian Gulf War between the United States and Iraq, which had invaded the neighboring country of Kuwait, as the Iraqi soldiers began retreating from Kuwait, they opened pipeline valves and set fire to oil wells. As a result, more than 895,000 tons (811,930 metric tons) of oil entered the Persian Gulf, polluting the marine ecosystem and decimating local wildlife.

Land-Based Oil Pollution

Oil in the ocean that originates on land is called nonpoint source pollution because it comes from many different sources rather than a single point of origin, such as a spill from a ship. Nonpoint sources can include runoff of oil and gasoline deposited on streets and parking lots by the millions of cars and trucks that travel a nation's roads. Industrial waste that may be contaminated with oil is also part of the land-based oil problem, as is wastewater from oil refineries. Improper disposal of petroleum products—including gasoline, oil, antifreeze, kerosene, and heating oil—can create another source of ocean pollution. In addition, the United States has nearly 23,000 miles (37,015 km) of petroleum pipeline, some of which is aging and vulnerable to leakage. Oil from these sources can eventually find its way to the oceans. Nonpoint source pollution is more harmful to the ocean environment than other sources, according to a report by the National Research Council, a science and public policy organization:

Oil spills can have long-lasting and devastating effects on the ocean environment, but we need to know more about damage caused by petroleum from land-based sources and small watercraft since they represent most of the oil leaked by human activities. . . . Oil slicks visible from the air and birds painted black by oil get the most public attention, but it is consumers of oil, not the ships that transport it, who are responsible for most of what finds its way into the ocean. For example, oil runoff from cars and trucks is increasing in coastal areas where the population is growing and roads and parking lots are expanding to accommodate it. Rivers polluted by oil in waste water or the improper disposal of petroleum products are a significant source of oil in the sea as well.[15]

Fouling Brazil's Beaches

Brazil's numerous and beautiful natural resources include the golden sand beaches that line its Atlantic coastline. But in late 2019 a 1,553-mile (2,500 km) stretch of these pristine shores became fouled with about 4,000 tons (3,630 metric tons) of crude oil probably spilled from an as-yet-unidentified tanker ship. The result has been the worst environmental crisis in Brazil's history.

Besides being a lucrative tourist attraction, Brazil's coastline is home to coral reefs and habitats for seabirds, fish, and humpback whales, all of which were and continue to be threatened by the oil spill. But the oil spill also affected the human inhabitants as well. Volunteer Mateus Morbeck, a local photographer who helped clean up the oil, commented on the process. "It is a horror and a disaster," he recalls, "and I found myself cleaning up the beach as I was taking photos. At first, we didn't know how to get rid of the oil. We didn't have any protection, no gloves, no masks, nothing. People were trying to do it with their bare hands and some began to faint and vomit. Eventually, we learned to protect ourselves."

The oil has also damaged Brazil's oldest marine national park, the Abrolhos Marine National Park, where several endangered species, including grouper fish and leatherback sea turtles, live. The spill on Brazil's coastline is a stark reminder of the ecological damage that pollution can cause.

Quoted in Mauro Pimentel, "Black Tide in Brazil," *AFP Correspondent* (blog), Agence France-Presse, November 14, 2019. https://correspondent.afp.com.

Watersheds, also known as catchments or drainage basins, are geographical areas that accumulate the flow of water from streams, rivers, lakes, and reservoirs and direct it eventually to the sea. According to NOAA, watersheds can be as small as a few acres or as large as thousands, or even millions, of square miles and are often hundreds of miles from the ocean. But when the rivers or streams that feed the watershed are contaminated with oil, it eventually ends up in the oceans. For example, in 2018 a slick of oil was spotted on the Mystic River in Massachusetts, which fed into a stormwater system and then flowed to the local watershed. Officials spent weeks searching for the source of the oil, finally identifying a leak in an underground pipe as the culprit. An estimated 6,870 gallons (26,006 L) of oil had leaked into the river, prompting a cleanup effort that lasted for months.

Oil and gasoline that leaks from the millions of cars traveling on highways eventually finds its way into the world's oceans.

Oil and the Environment

Oil spills, whether large or small, are harmful to the ocean ecosystem. In recent years there has been a reduction in the incidence of spills from oil tanker accidents. In the 1970s an estimated fifty to one hundred oil spills occurred each year; by the twenty-first century, that number had decreased to about twenty spills annually, thanks in large part to international agreements that require tankers to be double hulled (that is, have an interior and exterior hull). These numbers reflect only large spills of 7.7 tons (7 metric tons) or more, since smaller spills are seldom reported and thus more difficult to track.

The spills that do occur have a long-lasting effect on the environment. Oil can remain in the ocean for years or even decades, both on and beneath the surface of the water, as well as in coastal areas affected by a spill. For example, the oil released by the *Exxon Valdez* in Alaska in 1989 was still detectable more than ten years later. As another example, in 2019, fifty years after an oil barge spilled 175,000 gallons (662,447 L) off West Falmouth, Massachusetts, traces of the oil remained in the environment.

Oil spills are a complicated problem. According to the Woods Hole Oceanographic Institution, "Oil is not a single substance. . . . Crude oil and many refined petroleum products are a complex mixture of hundreds of chemicals, each one with a distinct set of behaviors and potential effects when released into the marine environment."[16] Because of this, each oil spill has its own characteristics and hazards, and thus each requires the development of unique ways to deal with the resulting pollution. The most serious consequences of oil spills are those that affect marine wildlife.

Wildlife at Risk

When oil spreads over the surface of the water, it prevents the absorption of oxygen from the atmosphere, depriving sea plants and animals of the oxygen they need to survive. Oil spills can also create dead zones, oxygen-depleted areas that can extend to the sea floor. The two main types of oil—light and heavy—create different kinds of hazards. These hazards are detailed by NOAA's Office of Response and Restoration, which responds to US coastal and environmental threats, including oil spills:

> Light oils present two significant hazards. First, some can ignite or explode. Second, many light oils, such as gasoline and diesel, are also considered to be toxic. They can kill animals or plants that they touch, and they also are dangerous to humans who breathe their fumes or get them on their skin.
>
> In contrast, very "heavy" oils (like bunker oils, which are used to fuel ships) look black and may be sticky for a time until they weather sufficiently, but even then they can persist in the environment for months or even years if not removed. . . . The short-term threat from heavy oils comes from their ability to smother organisms whereas over the long-term, some chronic health effects like tumors may result in some organisms.[17]

Oil that spills into the ocean is dangerous to marine mammals like this sea otter.

Tracking Arctic Oil Spills

Oil spills are difficult to track in the open ocean, but they are even more elusive if a spill occurs in the ice-covered Arctic seas. A new technology has been developed to help the US Department of Homeland Security (DHS) and the US Coast Guard prepare responses to oil spills in the Arctic. It involves deploying an underwater robot. "Because of ice coverage and the tyranny of distance," says Kirsten Trego, executive director of the US Coast Guard's Interagency Coordinating Committee on Oil Pollution Research, "it is difficult to get resources and assets up in the Arctic in a quick manner. With better real-time data, more effective response strategies can be developed and deployed." To help obtain that data, an underwater robot was developed by the DHS through its Arctic Domain Awareness Center, in cooperation with the Monterey Bay Aquarium Research Institute and the Woods Hole Oceanographic Institution.

The robot, known as a long-range autonomous underwater vehicle, or LRAUV, looks like a torpedo and is equipped with oil sensors. Launched from a ship or helicopter, the LRAUV can travel underwater (and under the Arctic ice), producing and transmitting 3-D maps of crude oil, diesel, gasoline, and kerosene spills. It can travel more than 370 miles (595 km) and remain operational for fifteen days before its batteries need recharging. With detailed oil-spill information provided by the LRAUV, workers can be more effective in cleaning up a challenging source of ocean pollution.

Quoted in US Department of Homeland Security, "Snapshot: More Tests for Arctic Oil-Spill-Mapping Robot," December 4, 2018. www.dhs.gov.

Sea otters are often referred to as a "keystone species" because of the beneficial effect they have on the marine ecosystem: sea otters prey on creatures such as sea urchins that disrupt the kelp ecosystem near shorelines. The International Union for Conservation of Nature has listed sea otters as an endangered species, in part due to the risks of oil spills. The sea otter's fur coat, which is one of the thickest in nature, serves as a protective barrier since the otter has no insulating fat. When oil, especially the heavy variety, coats the sea otter's fur, it loses its insulating capacity and leaves the otter vulnerable to the elements. This can result in death by hypothermia. Also, if a contaminated sea otter grooms itself, it can ingest toxins from the oil, causing kidney, liver, and lung damage. Oil may also damage an animal's reproductive system, leading to long-term effects such as deformities in offspring. Even the eggs of fish and birds can be harmed by oil.

As with otters, loss of insulating properties leading to hypothermia is just one of the effects that oil contamination has on seabirds. A dense coating of oil can make it difficult for birds to escape predators. Often prevented from flying due to the coating of oil, seabirds such as cormorants, gulls, petrels, terns, and other species may drown as a result of being weighed down by oil contamination. It is difficult to know exactly how many birds are affected by an oil spill. When these drownings occur out to sea, far from the nearest shoreline, they are rarely discovered or tallied. "A recent . . . experiment suggests that most birds oiled at sea sink there—never to wash up on a beach and be counted,"[18] notes seabird biologist Janet Russell.

Although there are fewer spills in the ocean today than there were decades ago, oil is still a major marine pollutant. Besides oil's short-term effects of injuring and killing wildlife, in the long term it can permanently change a marine habitat. As long as human society continues its dependence on oil, the chances of another oil-spill disaster happening are all too real.

Chapter Three

Chemical and Wastewater Pollution

Cruise ships are often called "floating cities" with populations as large as three thousand passengers. The largest ships, which are longer than three football fields, can accommodate more than six thousand travelers. Ocean cruising is a major part of the worldwide tourism industry, which had an estimated worth of $45.6 billion in 2018. Television commercials for major cruise lines show cheerful vacationers enjoying sun-drenched activities aboard the most luxurious ships afloat. But behind the scenes, cruising contributes to ocean pollution. Cruise ships, as well as other types of oceangoing vessels, have regularly used the seas as a dumping ground for sewage and other trash.

Pollution from Seagoing Vessels

Because the oceans are so immense, occupying about 71 percent of the earth's surface, people for centuries have thrown toxic waste, including sewage and chemicals, into them. After all, they may have reasoned, a few loads of noxious chemicals or human waste would certainly not have any effect on such a vast expanse of water. Today, however, people know better—or at least they should.

Cruise ships contribute to ocean pollution by dumping sewage and trash.

Perhaps surprisingly, in the twenty-first century it is legal to use the oceans as dumping grounds. A vessel may discharge waste that has been pulverized and disinfected in a shipboard processing unit within 3 nautical miles (5.6 km) of the nearest shore; untreated waste can be dumped farther than 12 nautical miles (22.2 km) from shore.

In 2019 the international environmental organization Friends of the Earth published a report describing the impact that cruise lines have on the environment. Using statistics from the US Environmental Protection Agency (EPA), the study reported that a single cruise ship carrying three thousand passengers can generate up to 150,000 gallons (567,812 L) of sewage each week. This amount adds up to a staggering 1 billion gallons (3.8 billion L) of waste discharged into the ocean every year across the entire cruise industry.

Cruise ships are not the only vessels that pollute the oceans. Cargo ships, tankers, and other seagoing merchant vessels add to the problem of ocean pollution. Transoceanic shipping accounts for more than 90 percent of all international commercial trade, creating a $3 trillion global economy. Since the early 1990s oceangoing traffic has increased dramatically. "In 20 years," says French oceanographer Jean Tournadre, "the growth is almost fourfold, or almost four times larger. We are putting much more pressure on the ocean."[19] Such a booming industry increases the chances of accidental, or deliberately illegal, pollution of the ocean environment. In addition to wastewater and sewage, other items that fall from cargo ships can be dangerous. In 2018 a cargo ship sank in the Persian Gulf off Dubai, releasing 18 tons (16.3 metric tons) of waste and goods. According to the Dubai Municipality, these goods included "electrical appliances such as refrigerators and air conditioners, which are dangerous to the marine environment."[20]

Wastewater discharged from oceangoing vessels generally falls into two categories: black water, composed of human waste; and gray water, which is wastewater from sources other than toilets, such as showers, sinks, baths, washing machines, and dishwashers. Other discharges from ships include bilgewater that drains from the deck to the bilge, or lowest part of the ship, and ballast, which is water held in tanks aboard ships to improve stability at sea. All these sources can damage the ocean environment. Black water contains disease-causing viruses and bacteria that can harm aquatic life and cause human diseases such as gastroenteritis, encephalitis, hepatitis, and typhoid fever. Gray water might contain remnants of chemicals from soaps and detergents as well as microbes that make it unsuitable for human consumption. Bilgewater is often tainted with oil, metals, and chemicals that are detrimental to the ocean ecosystem.

IMPACT FACTS

A typical cruise ship generates 150,000 gallons (567,812 L) of sewage every week—enough to fill ten standard-sized backyard swimming pools.

—Friends of the Earth

Land-Based Pollution

As detrimental as ships are for the ocean environment, they are only a relatively small portion of the entire ocean pollution picture. According to NOAA, the largest portion of chemical pollutants in the ocean—80 percent—comes from land-based sources. For countless years, people have been using the world's oceans as a garbage dump for unwanted chemical and other land-based waste. The Clean Water Act of 1972 helped reduce pollution in US waterways, but much damage had already been done. "Before 1972," says Jenny Howard, a journalist for *National Geographic* magazine, "humans around the word spewed trash, sewage sludge, and chemical, industrial, and radioactive wastes into the ocean with impunity."[21] Much of this pollution begins its journey far away from the oceans, as Howard explains:

> Many ocean pollutants are released into the environment far upstream from coastlines. Nutrient-packed fertilizers applied to farmland, for example, often end up in local streams and are eventually deposited into estuaries and bays. These excess nutrients trigger massive blooms of algae that rob the water of oxygen, leaving dead zones where few marine organisms can live. Some chemical pollutants climb high into the food webs—like DDT, the insecticide that placed the bald eagle on the United States Fish and Wildlife's endangered species list.[22]

In the United States some 34 billion gallons (129 billion L) of wastewater are treated every day. But the situation in developing countries is another story. Much of the land-based pollution that contaminates the oceans comes from underdeveloped countries

In 2011 an earthquake and tsunami struck Japan, damaging the Fukushima Daiichi Nuclear Power Plant and causing a massive meltdown of three of the plant's nuclear reactors. The disaster, the worst since a meltdown at a nuclear plant in Chernobyl, Ukraine, in 1986, spewed radioactive material into the air and the Pacific Ocean. After the disaster, workers began the long process of removing radioactive compounds from radiation-contaminated water, which had served to cool the now-destroyed reactors. The processed water is deposited in nearly one thousand storage tanks at the Fukushima site. By 2020, nearly one thousand tanks had been filled, with more water still to be processed and stored. As tank space is expected to reach maximum capacity by the middle of 2022, Japanese officials are considering another option: dumping the water into the Pacific Ocean.

While some of the radiation in the stored water has been removed, a radioactive substance called tritium remains. Although not directly harmful to humans, tritium harms fish and other aquatic life. Local fishers, whose fishing businesses were crippled by consumers' fears of radioactive fish, worry that their livelihoods could once more be disrupted.

In 2020 Japan's Ministry of Economy, Trade and Industry stated that dumping the stored water into the ocean is the only viable solution and that the water is not harmful. However, the environmental action group Greenpeace opposes the plan, claiming that continued storage of contaminated water is the safest solution. "The sea is not a garbage dump," says Greenpeace nuclear energy expert Jan Hakerkamp. "The sea is a common home for all people and creatures and must be protected."

Quoted in Aria Bendix, "Fukushima Is Running Out of Space to Store Contaminated Water. Japan's Environment Minister Said the 'Only Option' Is to Dump It in the Ocean," Business Insider, September 11, 2019. www.businessinsider.com.

around the world that either lack facilities for treating wastewater before it enters the oceans via streams and rivers or have facilities that are inadequate to treat the amount of wastewater produced. In fact, the United Nations reported in 2017 that in low-income countries, only 8 percent of wastewater receives any kind of treatment. In Africa there are more than three hundred urban centers along the continent's 18,950-mile (30,497 km) coastline. This concentration of people so close to the sea generates massive amounts of human waste and other pollutants (such as plastic and chemicals) that readily find their way into the oceans via rivers and other waterways.

Chemicals and Heavy Metals

Chemicals from households and industry can also end up polluting the oceans via rainwater runoff or improper disposal. A wide range of chemical substances contributes to ocean pollution: chemical waste produced by industry, including adhesives, flame retardants, oil, and solvents; and household products such as bleaches, detergents, lawn care products, and paints. These pollutants are known as persistent organic pollutants (POPs) and are harmful to humans and wildlife. An example of the harmful effect of POPs can be illustrated by their impact on the beluga whales

Beluga whales living in Canada's St. Lawrence Estuary have suffered an enormous decline due to poisoning by substances known as persistent organic pollutants (POPs).

of Canada's St. Lawrence Estuary. A population once numbered at about five thousand has been reduced to less than five hundred. In examining dead belugas, researchers found numerous physical disorders due to extensive POP pollution. Toxin levels in some of the St. Lawrence belugas are so high that these whales are considered toxic waste. This has a negative effect on the local Inuit people, for whom belugas have long been a dietary staple: cancer-causing toxins in beluga blubber has made the whale unsuitable for human consumption.

Heavy metals pose another threat to the ocean environment. Mercury is a toxic metal that occurs in nature, but it is also produced by industrial processes such as iron and steel production, mining, oil refining, and pulp and paper manufacturing. The burning of coal is also a major source of environmental mercury. It is well known that mercury is a health hazard to humans. Exposure to even small amounts of mercury can damage the heart, kidneys, lungs, and other vital organs, as well as have a serious impact on brain development in infants and children. The major cause of mercury exposure in the United States is the consumption of fish contaminated by the toxin. Mercury is ingested by the ocean creatures at the bottom of the food chain, such as zooplankton and mollusks, which are in turn eaten by small fish. As these creatures eat more and more mercury, poisons build up in their bodies through a process called bioaccumulation. In another process called biomagnification, the amount of toxins that are passed along to the small fishes' predators increases at each step in the food chain. Thus, the larger predator fish, such as halibut, mackerel, and tuna, have increased amounts of mercury in their systems, which ultimately presents a health hazard to humans who consume them.

Other metals such as aluminum, cadmium, cobalt, copper, lead, and zinc are all in varying degrees toxic to the ocean and its inhabitants. Aluminum, for example, inhibits respiratory functions in fish by accumulating on their gills, leading to breathing difficulties and death. It can also interfere with calcium regulation in their systems. Lead is a common metal used as fishing weights. These small weights can be eaten by seabirds and other marine life, leading to lead poisoning and often death. Dredging is another way that heavy metals can enter the oceans. When harbors are dredged in order to clean or expand them, toxic pollutants like zinc, cadmium, and other harmful metals can find their way into the oceans, threatening the viability of sea life.

Agricultural Pollutants

It takes many chemicals to grow crops, and these substances have become another form of ocean pollution. Farmers need fertilizer to enrich the soil, pesticides to ward off destructive insects, and herbicides to kill invasive weeds. When farmers plow their fields, the soil, along with the chemicals it contains, becomes exposed to the elements. Heavy rainfalls, especially in the spring, can wash these chemicals into watersheds and ultimately to the oceans, where they disrupt the ecosystem. Diana Papoulias, a fish biologist with the US Geological Survey, says that studies show how aquatic wildlife is affected by farm chemicals. "These aren't the kinds of studies that are done routinely, because they are pretty difficult to do," Papoulias explains. "But we know that some of these chemicals that we're finding in the runoff from the ag[ricultural] fields can affect [aquatic] reproduction and egg production."[23]

Agricultural nutrients such as nitrogen, phosphorous, and potassium play an important role in growing healthy and abundant crops. But when these chemicals enter the oceans, they can become too much of a good thing. A process called eutrophication occurs when an ocean environment becomes overloaded with too many nutrients. These nutrients usually enter the oceans via runoff from farmers' fields or from any area that requires fertilization: residential lawns, golf courses, public parks, and the like. The problems presented by excess nutrients in the ocean begin with organisms living at the very bottom of the food chain: phytoplankton.

Phytoplankton are microscopic algae that live in aquatic environments. There are an estimated five thousand species of

phytoplankton, and they feed on nutrients found in their aquatic environment. These tiny plants are important for the marine ecosystem because they serve as food for clams, oysters, scallops, and other shellfish, which in turn feed higher levels of the food chain. But through eutrophication, phytoplankton can ingest too many nutrients (mainly nitrogen and phosphorous), growing rapidly out of control and forming algal blooms. These blooms can color the surface of the water brown, green, or red and can grow so large that they can be photographed by satellites in orbit.

Some algal blooms are not dangerous, but others, called harmful algal blooms (HABs), can damage the marine ecosystem and its inhabitants. When blooms cover the surface of the water, they create oxygen-depleted dead zones, where organisms cannot live. One of the largest dead zones is in the Gulf of Mexico, where agricultural runoff from the farm-rich land of the American Midwest reaches the gulf via the Mississippi River. The zone is a 7,728-square-mile (20,015 sq. km) area where aquatic life cannot exist.

HABs can be toxic, killing fish, seabirds, turtles and other animals, and they can be harmful to humans as well. "If you go swimming in [a HAB], at the minimum your intestines will be in distress," says Richard Stumpf, a NOAA oceanographer. "If there's a lot, there's a risk of liver damage. Some people can get dermatitis from skin exposure to high concentrations of the toxin."[24]

Pollution Can Begin at Home

Even people who live far from an ocean can be responsible for some of the pollution that affects the aquatic environment. It typically takes several chemical products to keep a modern household clean and running. Detergents for cleaning clothes and dishes; paints, solvents, and varnishes; fertilizers and weed killers to keep lawns and gardens healthy; chemicals for swim-

The Tragedy of Minamata

One of the most horrific examples of the harm that toxic chemicals can do to humans happened nearly seventy years ago in the city of Minamata on the western coast of the island of Kyushu in Japan. In 1956 several people came down with a strange disease, with symptoms that included numbness of hands and feet, difficulty speaking or walking, weakness, and vision problems. Doctors were baffled as first four and then thirteen more people died of the mysterious disease.

Eventually, the cause of the disease was traced to a nearby chemical factory that had been dumping industrial waste into Minamata Bay for more than twenty years. Researchers discovered methylmercury, an extremely toxic form of mercury, in the factory's waste discharge. This chemical was absorbed by fish in the bay, which made up a large part of the local diet. People ate the fish and became poisoned by the mercury compound. Not only people but animals and birds became sick and died, including cats that lived in the factory and ate mercury-contaminated food.

Eventually, the factory ended the use of methylmercury in its manufacturing processes, and new cases of the disease stopped. But the damage had been done. According to the Japanese government, 2,955 people were certified as victims of mercury poisoning, and 1,784 of these have died, all due to the chemical pollution of Minamata Bay.

ming pools; and even personal cosmetics can all release toxins of various types.

Most manufacturers of household chemicals include instructions for proper disposal of their products. But not everyone follows these guidelines, instead disposing of the products by pouring them down the sink or flushing them down the toilet. These chemicals can go through a septic or municipal sewer system and ultimately be discharged into rivers and streams that empty into the ocean. Many of these hazardous chemicals pass through the municipal systems untreated, leaving them to potentially contaminate the water downstream.

Of the thousands of chemical substances used for household maintenance, many do not break down in the oceans and can build up toxins in fish and aquatic plants. And while some chemicals are relatively harmless by themselves, studies show that two or more may combine to create a dangerous mix in the environment. Pesticides are an example of such combining to create new and hazardous compounds. According to NOAA's Nathaniel Scholz:

> Current risk assessments based on a single chemical will likely underestimate impacts on wildlife in situations where that chemical interacts with other chemicals in the environment. The current findings may have implications for human health because these insecticides act on the nervous systems of salmon and humans in a similar way. Also, mixtures of pesticide residues can be common in the human food supply.[25]

By employing good practices in properly disposing of chemical waste and sewage, both industry and citizens can help keep toxins out of the ocean environment.

Chapter Four

Coral Reefs at Risk

The Great Barrier Reef is the largest coral network in the world. Located in the Coral Sea, the reef is a series of twenty-nine hundred individual reefs and nine hundred islands extending about 1,429 miles (2,300 km) along the coast of Queensland, Australia. What may surprise some people is that the Great Barrier Reef is alive. In fact, it is the largest living structure on earth and the only one visible from space.

Coral is a living organism, an invertebrate that clusters with neighboring coral to form vast networks that stretch for miles. The reef is its own ecosystem, supporting numerous forms of marine life. For example, the Great Barrier Reef is home to more than fifteen hundred species of fish, including angelfish, perch, rays, sharks, trout, and numerous others. Dolphins, turtles, mollusks, whales, and even seabirds rely on the reef for food and shelter. But in the twenty-first century, the Great Barrier Reef is dying. After an exploratory dive to the reef in 2017, research scientist Neal Cantin surfaced and gave a concise, ominous report: "All dead."[26]

According to the National Geographic Society, about half of the coral has died since 2016. A 2019 report by the Australian government has officially downgraded the outlook of the reef from "poor" to "very poor," and some sources say that it could be gone by 2050.

What is happening to the Great Barrier Reef is also happening to coral reefs around the world. More than half of the world's coral reefs have died in the past thirty years, and 90 percent may die over the next one hundred years. And pollution is playing a part in the destruction of these vital aquatic habitats. If the increasing death of coral reefs is not reversed, says J.E.N. Veron, a marine researcher and coral expert, the world "will eventually see a meltdown in coastal economies with devastating cost to natural environments and human societies."[27]

The Importance of Coral Reefs

The corals that form reefs are called polyps. There are two types of polyps, hard (also called stony) and soft. The hard polyps are the coral reef builders. They live in a small, hard cup made of calcium carbonate, the main ingredient in limestone. This cup protects them from predators and allows them to grow. These

Australia's Great Barrier Reef, along with other coral reefs around the world, is dying, due in part to ocean pollution.

polyps create reefs by attaching their limestone cup to an under-water surface; millions of polyps attach more cups, and layers build up, eventually forming a large reef.

Although coral reefs occupy less than 1 percent of the world's oceans, their importance far outweighs their size. Coral reefs pro-vide a habitat for more than 25 percent of all known ocean life. This huge biodiversity is even larger than that of a rain forest—coral reefs are often called "rain forests of the sea"—and helps keep the ocean ecosystem clean and better able to withstand en-vironmental changes. It also helps provide food for people living in lands near a reef, employment through the fishing and tourism in-dustries, and protection for coastlines against floods, storms, and waves. Reefs even make contributions to health care through the chemical compounds they produce to defend themselves from

Coral Reefs and Noise

Snorkeling or scuba diving can be a tranquil and uplifting experience for vaca-tioners exploring the coral reefs in the clear waters of the Caribbean and other tropical environments. But for the reefs it is a noisy environment, and it is the sounds that various species of fish make that helps the reefs thrive. Unfortu-nately, there are other kinds of noise that can have a damaging effect on the delicate corals of the world.

The idea that natural noise can help coral reefs stay healthy has been test-ed by ocean scientists, who played recordings of healthy corals near dead and dying reefs. The scientists discovered that more fish were attracted to the dead corals simply because they sounded healthy. But other sounds are not so ben-eficial to the reefs. Just as human activity on dry land can be a cacophony, so it can beneath the waves. Loud sounds from boats (especially recreational boats that operate near shorelines), underwater mining, and other noise-producing activities can put stress on fish and other sea creatures that healthy reefs de-pend on. Noise pollution can also affect coral larvae, or "baby" corals. In their first days of life, these larvae must find a suitable habitat to attach to and grow by homing in on the natural noise of corals. When anthropogenic noise masks the natural sounds of the reef, the larvae become disoriented and cannot find a safe place to live and grow.

predators. Researchers study these compounds for their medicinal benefit to humans, according to coral reef ecologist Andrew W. Bruckner:

> The antiviral drugs Ara-A and AZT and the anticancer agent Ara-C, developed from extracts of sponges found on a Caribbean reef, were among the earliest modern medicines obtained from coral reefs. Other products . . . are under clinical trials for use in the treatment of breast and liver cancers, tumors, and leukemia. Indeed, coral reefs represent an important and as yet largely untapped source of natural products with enormous potential as pharmaceuticals.[28]

While coral reefs provide numerous benefits for animals and humans, their delicate nature makes them vulnerable to environmental assaults by ocean pollution.

Mystery at Looe Key

In the Florida Straits about 25 miles (40.2 km) from Key West, Florida, lies Looe (pronounced "Loo") Key, a reef that is part of the Florida Keys National Marine Sanctuary. Managed by NOAA, the 3,840-square-mile (9,946 sq. km) sanctuary protects some seventeen hundred islands and more than six thousand marine species. It is the only barrier reef in North America.

A 2019 report of a study that collected data over thirty years announced an alarming decrease in the coral of the reef. During the three decades studied, coral coverage of Looe Key decreased from 33 percent to just 5 percent. Something was killing the Looe Key coral. "Watching the decline of coral at Looe Key has been heartbreaking," says Brian Lapointe, a research professor at Florida Atlantic University's Harbor Branch Oceanographic Institute and lead author of the 2019 report. "When I moved here in the early 1980s, I had no idea that we would be losing these corals."[29]

Reef-building corals can survive only in warm water, which is why they are found in shallow tropical and subtropical environments. The range of water temperatures in which these corals thrive varies from about 68°F to 90°F (20°C to 32°C). Ocean temperatures falling above or below this range will put corals at risk. But it is not just the water temperature that can have a devastating effect on coral, and Lapointe soon discovered the cause of the Looe Key coral deaths: an increase in nitrogen in the water and a decrease in phosphorous, an element needed for coral growth. "The corals are assimilating nutrients from the water column," he says. "This [nitrogen to phosphorus imbalance] is what we now realize is increasingly stressing the corals at Looe Key, and probably other areas of Florida and the world."[30]

The Death of Coral Reefs

Looe Key, like all coral reefs, is a strikingly beautiful structure when it is healthy. The calcium carbonate skeletal structure of coral itself is white. Coral's color comes from a type of algae called zooxan-

Stress caused by exposure to ocean pollution causes coral to expel the microorganisms that give it color, a process known as coral bleaching.

thellae. These tiny organisms live on coral in a symbiotic, or mutually beneficial, relationship. The zooxanthellae create oxygen and nutrients needed to keep the coral alive, while the coral in turn protects the algae from predators and helps with photosynthesis.

It is the zooxanthellae that give coral its shades of blue, green, red, violet, and other vibrant colors. These colors, besides giving snorkeling tourists a beautiful scene to admire, are also an indication of the health of a coral reef. Corals and their zooxanthellae partners can suffer from environmental stress, altering their symbiotic relationship. When the stress becomes too much, the coral will expel the zooxanthellae, leaving behind the reef's white skeleton in a process known as coral bleaching. Deprived of its colorful companion and the vital nutrients it provides, the bleached coral soon dies. Mass coral bleaching events regularly occur; one such event lasted from 2014 to 2017, affecting more than 75 percent of all coral reefs on the planet.

IMPACT FACTS

Two kinds of reef-building corals, elkhorn and staghorn, have declined 92 to 97 percent since the 1970s.

—Ocean Conservancy

Ocean pollution plays a large part in creating the stresses that promote the destruction of coral reefs. In the case of Looe Key, the increase of nitrogen in the water caused by excessive agricultural runoff is the primary source of reef-killing stresses. Heavy rains, made more frequent by climate change, washed agricultural fertilizer containing nitrogen into the ocean environment where the coral lives. What happened at Looe Key is happening all over the world.

How Construction and Industrial Pollution Harm Coral Reefs

Many types of anthropogenic pollution can affect the delicate structure of coral reefs. As coastal populations expand, increases in pollution are inevitable. For example, the growing construction of roads, commercial buildings, and residences in coastal areas around the world creates an excess of dirt, sand, and other by-products of construction. Rains wash this debris into the ocean, resulting in increased sedimentation, which is the tendency of these particles to accumulate at the bottom of a body of water. Since most coral reefs reside in shallow areas along shorelines, sedimentation is a threat to their health. Sediment can cloud the water in these areas, depriving corals and zooxanthellae of the sunlight they need to survive. Reefs can be smothered by too much sediment covering them, which weakens them as they expend more energy to clean themselves of debris.

Coastal construction activities often include removal of mangrove trees. Mangroves grow near the shore in tropical areas and act as a barrier against coastal erosion and sedimentation. Without these trees acting as a filter, sediment can enter the water more easily to affect local coral reefs.

Industrial chemicals can deprive coral reefs of necessary elements for survival. When discharged into the ocean, these chemicals encourage the formation of algal blooms, which cover the surface of the water and reduce the amount of sunlight and oxygen coral reefs need. In one instance an algal bloom with an area of 193 square miles (500 sq. km) developed in the Gulf of Oman, an arm of the Arabian Sea between the nations of Oman and Iran. When researchers from the United Nations University Institute for Water, Environment and Health studied coral reefs in the area, they discovered that within three weeks of the formation of the bloom, the reefs had almost completely died off. Fish that relied on the coral reefs for their habitat were also severely affected, with species such as goatfish, parrotfish, and snapper diminished or eliminated. "We were surprised at the extent and speed at which changes to the coral reef communities were affected,"[31] notes Andrew Bauman, a marine ecologist at the University of Singapore.

The Threat of Sewage

While chemicals and plastics are harmful to coral reefs, sewage discharged into the oceans poses another threat. In poor nations, where treatment of human waste is inadequate or nonexistent, sewage is the most widespread form of water pollution. "Some of the greatest sewage pollution in the world occurs in developing countries, where you also have the most coral reefs,"[32] says Stephanie Wear, a marine ecologist and coral conservation expert with the Nature Conservancy. Wear's studies have revealed that 96 percent of locations where there are people and coral reefs are contaminated with sewage pollution.

Sewage contains numerous compounds that are deadly to corals: bacteria and viruses; heavy metals; endocrine disruptors, which affect hormone systems; and synthetic toxins, including antibiotics and other medicines. Kaneohe Bay lies on the eastern coast of Oahu in Hawaii, and from the 1930s to the 1970s it

Sewage dumped into oceans contains numerous compounds and organisms that are deadly to coral.

Plastic and Corals

Plastic, the most common ocean pollutant, not only fouls the marine environment and causes the deaths of some 100 million sea creatures every year, it also kills coral reefs. Randi Rotjan is a biology professor at Boston University, and while researching coral reefs, she discovered microplastics in virtually all the coral samples she studied. Rotjan found that, like people, corals have an immune system that can sometimes be vulnerable to infections:

> If you feed corals plastics, and the plastics have bacteria on the surface, and they're harmless bacteria and the coral's immune system is intact, it's probably not going to be that big a deal in terms of microbial infection. But if you have an infectious pathogen and a coral's immune system is suppressed, and microplastics act as a little plastic raft where all these bacteria can hitch a ride, then suddenly maybe these plastics are not just plastics. They're plastics with a problem attached.

Quoted in Barbara Moran, "Researchers Find Another Threat for Corals: Plastic," WBUR-FM (Boston), June 26, 2019. www.wbur.org.

was contaminated with sewage that killed some 90 percent of the bay's reef system. As soon as the sewage was diverted in 1978, however, the coral began to regrow.

Perhaps most surprising, freshwater is deadly to coral, because it upsets the balance of salt in the salt water that coral requires for survival. In 2014 an influx of freshwater at Kaneohe Bay reduced the coral in the bay by more than 22 percent. This "freshwater kill," combined with a subsequent bleaching of the coral due to rising water temperatures, killed more than 60 percent of the coral in the area. But threats to coral reefs are found not only in water but in the atmosphere itself in the form of air pollution.

Climate Change and Coral Reefs

Ever since the beginning of the Industrial Revolution in the mid-eighteenth century, the earth's climate has been slowly changing. Factories, motor vehicles, power plants, airplanes, household chemicals, and landfills create pollution that remains in the atmosphere. Some of these pollutants are made up of very fine particles; the harmful effects on air quality of this particulate matter also impact the ocean environment and the plants and animals living there. It is a global problem, and an economic one, since coral reefs are a major tourist attraction in many parts of the world. As Ian McPhail, a professor at the University of Southern Queensland in Australia and a former chair of the Great Barrier Reef Marine Park Authority, explains:

> When it comes to looking after coral reefs, we tend to sweat over local issues like port development, dredging, nutrient flow from the land and so on. They're all very important—but this is a reminder of some of the bigger regional and global atmospheric challenges we need to address too. . . . We must remember, we're not just talking about the ecology of coral reefs, we're also talking about the economy.[33]

When these pollutants enter the atmosphere, they form a barrier that reduces the amount of sunlight that reaches the earth below. This prevents the photosynthesis by which the zooxanthellae make the nutrients and oxygen that the corals need to grow and reproduce. Atmospheric pollutants also trap heat that would otherwise be reflected out into space, creating a steadily warmer climate. The oceans receive most of that heat energy— according to NOAA, they have absorbed more than 90 percent of all the global warming in the past fifty years. Increased levels of carbon dioxide in the atmosphere are also absorbed by the oceans, making the seawater more acidic, which weakens the corals' calcium carbonate skeleton. Coral bleaching follows, often with fatal results.

"Coral reefs have for a long time been considered to be . . . the canary in the coal mine, which foretells substantial and deleterious changes happening to the Earth system,"[34] says ecology professor Charles Sheppard. In the early days of coal mining, a canary in a cage was taken into the mines. As long as the canary sang, miners knew that the underground air was safe. But if the singing stopped, it was a warning that toxic air was present and the mine should be evacuated. Coral reefs are the marine version of the canary in the mine; their bleached skeletons provide an early warning that toxic pollution is destroying a vital part of the earth's environment.

Chapter Five

Searching for Solutions

Given the enormous expanse of the earth's oceans and the continuing human activity impacting them, confronting the problem of ocean pollution may seem a hopeless undertaking. But many people are beginning to realize the consequences of ocean pollution and are working to clean up the seas. "To be so ignorant and neglectful of our oceans is deeply troubling," says oceanographer Sylvia Earle. "However, the optimist in me tells me that, having woken up to this living disaster and having realized that there are limits to how much abuse we can inflict, it's not too late to turn things around."[35]

Removing Microplastics

Plastic, the largest source of ocean pollution, may be the most challenging to clean up. The immense range in size of plastic in the oceans—from tiny microplastics to miles-long fishing nets—increases the difficulty of remedying the situation. But difficult tasks often call for unique solutions. And for one Irish teenager, an innovative idea for removing microplastics from the oceans won him a $50,000 prize.

Fionn Ferreira lives in West Cork, Ireland, at the bottom of the Emerald Isle on the Atlantic Ocean. One day on a kayaking trip, Ferreira noticed many small particles of plastic that were stuck to an oil-covered rock. Due to his interest in science, he knew that plastic tended to be attracted to oil. That led him to start thinking about the ocean and the pollution that befouled it. He decided to find out whether there was some way to remove plastic pollution from water.

In his early experiments, Ferreira placed vegetable oil into a vial of water containing microplastics. As he had hypothesized, the plastic particles adhered to the oil. But he still needed to find a way to remove the oil from the water. After much research, Ferreira learned about a substance called ferrofluid, a magnetic, oil-based liquid developed by the National Aeronautics and Space Administration and found today in consumer electronics. By adding magnetite, a naturally occurring magnetic mineral, to the vegetable oil, he made his own version of ferrofluid. Using this magnetic oil in his experiments, he was able to remove it, and the microplastics that were stuck to it, by using an ordinary magnet.

Ferreira performed hundreds of experiments using different types of plastic particles, and he built homemade instruments to measure his results. His method removed some 87 percent of plastic microparticles from water, a much higher amount than that obtained by ordinary wastewater filtration methods and slightly more than his predicted results of 85 percent. For his unique solution to the microplastics problem, Ferreira won the $50,000 grand prize in the 2019 Google Science Fair. In 2019 he was a college student in the Netherlands, where he worked on scaling up his project to make it a practical solution to plastic pollution.

Scooping Up Plastics

While Ferreira's inventive method may one day help rid the oceans of microplastics, another young engineer has plans to clean up the GPGP. Like Ferreira, sixteen-year-old Boyan Slat had an encounter with plastic in the ocean that inspired him to devise a way to clean up ocean pollution. In 2012, after experimenting with various concepts and testing his theories, Slat came up with a way to use ocean currents to gather plastic pollution so it could be collected and recycled. The next year he founded the Ocean Cleanup, an organization dedicated to putting his ideas

into practice. By 2014 the organization had raised over $2 million in crowdfunding campaigns.

Slat's idea employs a U-shaped floating plastic barrier that is towed out into the sea and, through the action of wind and ocean currents, collects plastic debris floating on the ocean's surface and several yards (meters) below it. A service vessel, acting as a sort of garbage truck of the ocean, periodically collects the plastic debris and transports it to shore to be recycled. In October 2019 the floating barrier, designated System 001/B, began real-world testing by collecting plastic debris—including large objects, ghost nets, and microplastics—from the GPGP. Slat envisions the eventual deployment of some sixty systems, which could clean 50 percent of the GPGP's plastic every five years.

Removing Oil Pollution

Oil in the ocean is notoriously difficult to clean up, and many technologies are being used with varying degrees of success. One method uses floating booms, similar to the Ocean Cleanup's plastic-trapping barriers. These devices, called containment booms, are deployed after an oil spill in order to keep the oil from contaminating nearby shorelines. Made of plastic, metal, or other materials, containment booms are towed into place around a spill and, depending on their construction, either surround the slick or absorb the floating oil.

Authorities used booms and other methods when responding to the massive *Deepwater Horizon* explosion and oil spill in 2010. The spill released 200 million gallons (757 million L) of oil into the Gulf of Mexico, damaging the fragile ecology of the Gulf Coast. BP engineers devised plans and built equipment over a span of several weeks to combat the spreading pollution.

Using satellite imagery and video-equipped unmanned submersibles, engineers calculated the extent and depth of the oil spill. To contain the oil, workers deployed 4.2 million feet (1.3 million m) of containment booms; 9.1 million feet (2.7 million m) of absorbent booms were also used to soak up the oil. Some of the oil that was enclosed by the containment booms was burned off to remove it from the surface of the water. Working in coordination with the US Coast Guard, BP carried out more than four hundred controlled burns, which removed 265,000 barrels of oil from the spill.

IMPACT FACTS

To contain the oil released from the *Deepwater Horizon*, workers deployed 4.2 million feet (1.3 million m) of containment booms.

—National Ocean Industries Association

BP used other techniques to clean up the spilled oil as well. The company used skimmers—some on floating booms, others attached to vessels—to remove the oil from the surface and deposit it in holding tanks. Aircraft sprayed chemical dispersants to break up the oil, allowing waves and bacteria to dissolve it. BP also constructed artificial barriers to protect the shore from oil damage. Five years of cleanup activities cost the company over $14 billion, not including billions more spent to cover federal, state, and local assistance.

Many regions of the oceans around the world have been designated as marine protected areas, or MPAs, which safeguard the vital natural resources of the seas. There are numerous types of MPAs; they can be as small as a local estuary or as large as the Antarctic MPA, which covers 600,000 square miles (1.6 million sq. km). The United States alone has more than seventeen hundred MPAs, including many coastal waters and the Great Lakes. According to oceanographer Sylvia Earle, "Marine Protected Areas help recover marine environments in order to provide natural solutions to critical environmental challenges. They provide safe havens for ocean wildlife to recover and maintain healthy biodiversity, supporting habitats that act as carbon sinks, removing CO_2 [carbon dioxide] from the air, and generating the majority of atmospheric oxygen."

MPAs can include marine nature reserves, national monuments or cultural sites, species management areas, and national parks. In many MPAs human activities such as boating, fishing, oil drilling, and tourism are restricted or prohibited. Individual MPAs can be connected to form networks designed to combine resources to meet similar ecological and socioeconomic goals.

Although the MPAs are intended to maintain a protective marine environment, pollution is encroaching into many of the areas. While information on pollution in MPAs is scarce, a 2018 study found that 80 percent of the protected areas examined showed signs of potentially toxic chemical pollution.

Quoted in David de Rothschild, *Plastiki: Across the Pacific on Plastic: An Adventure to Save Our Oceans.* San Francisco: Chronicle Books, 2011, p. 109.

Preventing Oil Pollution

While new technology has helped in cleaning up oil spills, the ocean environment would naturally be better off if oil were not spilled in the first place. Prevention is a much more desirable solution than going through the labor and expense of cleanup. Owners of oil tanker ships and oil refineries are making progress in using preventive measures to keep oil from spilling. Since 2015 ships that carry thousands of gallons of oil in US waters have been required to be constructed with double hulls. This will prevent spills should these vessels strike another ship or run aground. Likewise,

all oil-carrying barges that ply the rivers within the United States must also be equipped with double hulls.

Like double hulls, redundancy, or the use of duplicate equipment, helps make modern tankers safer. Tankers include redundancies such as backup Global Positioning System devices and radar navigation equipment and duplicate propulsion systems, including fully independent engine rooms. This assures that, if a vital system failure threatens the vessel, the crew can still maintain control. And that crew is made up of licensed professionals who have received practical ship-handling experience at training facilities that use virtual reality simulators and full-scale mock-ups of a tanker's bridge.

Cleaning Up Sewage and Chemicals

In addition to oil, approximately 2 million tons (1.8 million metric tons) of sewage enter the world's oceans every day. In many countries this waste is either treated inadequately or not treated at all. A Swedish company called SurfCleaner has developed a simple device to remove sludge from water in waste treatment plants before it ever gets to the oceans. The product, also called SurfCleaner, was inspired by a Swedish doctor's study of the way the human heart pumps. SurfCleaner is a simple surface skimmer that has only two moving parts. It separates sludge from the wastewater treatment plants and discharges it into a storage tank. The device is powered by solar cells and can run twenty-four hours a day, seven days a week, and it can separate about 2,113 gallons (8,000 L) of waste per hour. SurfCleaner is also used to remove oil and plastics from ocean surfaces.

Upgrading water treatment infrastructure helps reduce water pollution caused by human waste at the source. According to the World Bank, about 2.3 billion people around the world do not have access to basic sanitation services. The World Bank provides loans to fund sanitation improvement projects in poor nations around the globe. In Ghana, for example, the World Bank provides funding for families to purchase toilets and for the construction of new wastewater treatment plants.

Saving Coral Reefs

Efforts to clean up the pollution that is endangering ocean environments have given new hope to those who fight to save coral reefs. "We created these problems," notes Michael Crosby, president of Mote Marine Laboratory & Aquarium in Sarasota, Florida. "We have to get actively involved in helping the corals come back."[36]

Based in Key Largo, Florida, the Coral Restoration Foundation (CRF) is the world's largest nonprofit organization dedicated to the study and rebuilding of coral reefs. Working with ocean scientists, academic institutions, and coral experts, the CRF is rebuilding the world's coral reefs, the prime objective of its founder, Ken Nedimyer. While diving near where he grew up in the Florida Keys, young Nedimyer became fascinated by the beautiful colors and vibrant marine life of the coral reefs. But over time, he watched as the reefs slowly began dying. "I did a lot of thinking," Nedimyer says, "and it became a consuming passion to try to find ways to restore coral reefs."[37]

His idea was to grow new corals by creating underwater coral farms, or nurseries. The CRF trains volunteers who perform the actual underwater work of building a new reef. Divers first erect "trees" made from PVC pipe. Spindly arms sprouting from the trees give them the appearance of bare, white Christmas trees. Next, finger-sized pieces of living coral—called "tough coral" for their endurance through disease and bleaching—are tied to the trees' arms, usually about 150 per tree. Then comes the waiting as the coral grows. The coral pieces, says Nedimyer, "will hang here for about six months to nine months. After it's been here for about nine months, then we take it out onto the reef and we plant it."[38] By 2019 Nedimyer's organization had planted more than sixty-six thousand new corals in the Florida Reef Tract.

The Coral Restoration Foundation trains volunteers to plant tiny pieces of coral that are resistant to disease and bleaching, in an effort to build new reefs.

The successes of the CRF demonstrate that someone with a fresh idea can help improve the ocean's ecosystem. But Nedimyer knows that the job is bigger than just one person. "This isn't just about me," he says. "It's about engaging a lot of people and training people, and I think it has a lot of hope. I'm convinced this is the solution that can work."[39]

Legislating for Clean Oceans

In 1972 the US Congress enacted the Clean Water Act (CWA), also known as the Federal Water Pollution Control Act. It was one of the first legislative actions designed to prevent pollution of America's waterways. The CWA established the basic structure for regulating pollutant discharges into the nation's rivers, streams, lakes, and other bodies of water. After a 1977 amendment, the act included untreated wastewater from ships, factories, and municipalities. According to the EPA's website, acting under the authority of the CWA, the EPA "has implemented pollution control programs such as setting wastewater standards for industry. The EPA has also developed national water quality criteria recommendations for pollutants in surface waters."[40] The EPA was also given responsibility for issuing permits for discharges into the ocean. This requires the agency to assess the effect of proposed discharges on sensitive biological communities and aesthetic, recreational, and economic values before a permit can be issued.

America has cleaner lakes, rivers, and streams as a result of the CWA's pollution guidelines. But despite these improvements, the CWA is not immune to twenty-first-century politics. In 2020 President Donald Trump announced a rule that would reduce EPA protection for certain categories of US streams and wetlands. While the rule would likely benefit certain industries such as construction by easing dumping regulations, environmentalists foresee a rise in water pollution. "If you swim in a river or a stream," says Brett Hartl of the nonprofit Center for Biological Diversity, "or if you go fishing and you want to eat the fish you catch, that may have more toxic heavy metals in it than before."[41] At the time the

Mr. Trash Wheel

It looks like a gigantic bathtub toy, with its huge googly eyes staring out from a shell-like body. But this "toy," whimsically named Mr. Trash Wheel, has a serious purpose: cleaning trash from Baltimore's Inner Harbor before it can escape to the ocean.

The brainchild of John Kellett, Mr. Trash Wheel features a 14-foot (4.3 m) waterwheel that runs a conveyor belt to scoop up floating trash and deposit it into a dumpster. The wheel is powered by the harbor's current, or by solar cells if the river's pace is slow. Since its installation in 2014, Mr. Trash Wheel has scooped up more than 1,356 tons (1,230 metric tons) of trash, including almost 12 million cigarette butts, 1 million plastic bottles, 700,000 plastic bags—plus a guitar and a live python.

Mr. Trash Wheel has become a beloved part of Baltimore's waterfront and, perhaps not surprisingly, a tourist attraction. It (or is it he?) has a Twitter account and has spawned items such as tote bags, caps, and T-shirts. The Trash Wheel family has grown with the addition of two "cousins," Professor Trash Wheel in 2016 and Captain Trash Wheel in 2018. "We feel they're successful," notes Kellett, "both in what they accumulate in the tons of trash they pick up, and they're also very important from the standpoint of they draw people's attention to the problem, educate them about the problem and inspire them to become a part of the solution."

Quoted in Stephen Babcock, "The Company Behind Mr. Trash Wheel Is Working with Yamaha on a New Device in Georgia," Technical.ly, June 28, 2019. www.technical.ly.

rule was finalized on January 23, 2020, court challenges were expected to delay its implementation.

The year after the CWA was enacted, the most important global agreement to limit ocean pollution was adopted. In 1973 the International Convention for the Prevention of Pollution from Ships, known as MARPOL (for Marine Pollution), was introduced by the International Maritime Organization, an agency of the United Nations. According to the agency's website, MARPOL is the "main international convention covering prevention of pollution of the marine environment by ships from operational or accidental causes."[42] As of 2020, 174 nations, accounting for more than 97 percent of the world's shipping tonnage, have adopted the convention. MARPOL includes six annexes, or sections, establishing regulations for various types of marine pollution: oil, poisonous bulk liquids, harmful packaged substances, sewage, garbage (including plastics), and air pollution.

Although established almost half a century ago, MARPOL is still being amended to establish new goals in reducing ocean pollution. In 2020 a new provision was added requiring ships to limit sulfur emissions, which cause air pollution that eventually falls into the oceans as acid rain. Increasing the acidity of the ocean puts coral reefs, many species of fish, and marine animals that have shells, such as oysters and sea snails, at risk.

IMPACT FACTS

By 2018, 127 countries had enacted policies regulating plastic bags, including 32 that imposed outright bans.

—Business Insider

Fighting ocean pollution is neither easy nor inexpensive. To comply with the new MARPOL regulations, for example, many shipping companies will have to invest in new engine exhaust scrubbers or begin using more expensive, low-sulfur fuel. All this means more-expensive trips for cruise passengers and higher prices for goods shipped by sea. But given the importance of the world's oceans, it is an expense that humans must endure. In the words of the late marine explorer and ocean conservationist Jacques-Yves Cousteau, "The very survival of the human species depends upon the maintenance of an ocean clean and alive, spreading all around the world. The ocean is our planet's life belt."[43]

Source Notes

Introduction: The Endangered Oceans

1. Quoted in Alanna Mitchell, "Is That My Plastic Bag in the Mariana Trench?," *Maclean's*, July 2019, p. 76.
2. Quoted in Quotetab, "Sylvia Earle Quotations," 2019. www.quotetab.com.
3. Sylvia Earle, "Can We Stop Killing Our Oceans Now, Please?," HuffPost, December 6, 2017. www.huffpost.com.

Chapter One: Plastic Pollution

4. Quoted in Casey Conley, "Volvo Ocean Racers Gather Data on Plastic Pollution," *Ocean Navigator*, August 31, 2018. www.oceannavigator.com.
5. Quoted in Steve Toloken, "WWF: Plastic Pollution to Double by 2030," Plastics News Europe, March 11, 2019. www.plasticsnewseurope.com.
6. Quoted in Natural Resources Defense Council, "NRDC Lauds Passage of New York City Council Legislation Requiring Groceries, Retailers to Provide Plastic Bag Recycling for Consumers," January 9, 2008. www.nrdc.org.
7. Quoted in Helmholtz Centre for Environmental Research, "Rivers Carry Plastic Debris into the Sea," ScienceDaily, October 17, 2017. www.sciencedaily.com.
8. Quoted in Aylin Woodward, "New Research Suggests We Might Be Thinking About the Ocean Plastic Problem All Wrong—Trash Dumped from Ships Could Be a Major Culprit," Business Insider, October 4, 2019. www.businessinsider.com.
9. Charles Moore with Cassandra Phillips, *Plastic Ocean: How a Sea Captain's Chance Discovery Launched a Determined Quest to Save the Oceans*. New York: Avery, 2011, p. 4.
10. Megan Forbes, "Garbage Patches: How Gyres Take Our Trash Out to Sea," *NOAA Ocean Podcast*, National Ocean Service, March 22, 2018. www.oceanservice.noaa.gov.
11. Daniel Victor, "1,000 Pieces of Plastic Found Inside Dead Whale in Indonesia," *New York Times*, November 21, 2018. www.nytimes.com.

12. Lateshia Beachum, "Dead Sperm Whale Had 220 Pounds of Garbage in Its Stomach, Including Rope, Plastic and Gloves," *Washington Post*, December 2, 2019. www.washingtonpost.com.

13. Quoted in Victor, "1,000 Pieces of Plastic Found Inside Dead Whale in Indonesia."

Chapter Two: Oil Pollution

14. Quoted in Wallace White, "Her Deepness," *New Yorker*, July 27, 2014. www.newyorker.com.

15. National Research Council, "Millions of Gallons of Petroleum from Human Activities Enter North American Waters Annually; Most Comes from Runoff, Small Watercraft," 2002. www.nationalacademies.org.

16. Woods Hole Oceanographic Institution, "Oil in the Ocean: A Complex Mix," May 2, 2011. www.whoi.edu.

17. Office of Response and Restoration, "How Oil Harms Animals and Plants in Marine Environments," 2020. www.response.restoration.noaa.gov.

18. Janet Russell, "Untold Seabird Mortality Due to Marine Oil Pollution," Elements, October 2000. www.elements.nb.ca.

Chapter Three: Chemical and Wastewater Pollution

19. Quoted in Becky Oskin, "Ship Traffic Increases Dramatically, to Oceans' Detriment," Live Science, November 18, 2014. www.livescience.com.

20. Quoted in Nawal al Ramahi, "Commercial Ships Dump Waste into Sea, Say Maritime Officials," *The National*, March 29, 2018. www.thenational.ae.

21. Jenny Howard, "Marine Pollution, Explained," *National Geographic*, August 2, 2019. www.nationalgeographic.com.

22. Howard, "Marine Pollution, Explained."

23. Quoted in Abbie Fentress Swanson, "What Is Farm Runoff Doing to the Water? Scientists Wade In," NPR, July 5, 2013. www.npr.org.

24. Quoted in Mia Tramz, "See Striking New Images of Algae Blooms in the Great Lakes," *Time*, August 6, 2015. www.time.com.

25. Quoted in SeaWeb, "Chemicals in Our Waters Are Affecting Humans and Aquatic Life in Unanticipated Ways," Science-Daily, February 21, 2008. www.sciencedaily.com.

Chapter Four: Coral Reefs at Risk

26. Quoted in Damien Cave and Justin Gillis, "Building a Better Coral Reef," *New York Times*, September 20, 2017. www.nytimes.com.
27. J.E.N. Veron, *A Reef in Time: The Great Barrier Reef from Beginning to End*. Cambridge, MA: Belknap, 2008, p. 231.
28. Andrew W. Bruckner, "Life-Saving Products from Coral Reefs," *Issues in Science and Technology*, Spring 2002. www.issues.org.
29. Quoted in Pien Huang, "Florida's Corals Are Dying Off, but It's Not All Due to Climate Change, Study Says," NPR, July 16, 2019. www.npr.org.
30. Quoted in Huang, "Florida's Corals Are Dying Off, but It's Not All Due to Climate Change, Study Says."
31. Quoted in Ella Davies, "Toxic Algae Rapidly Kills Coral," BBC, October 8, 2010. http://news.bbc.co.uk.
32. Quoted in Justine E. Hausheer, "Sewage Pollution: A Significant Threat to Coral Reefs," *Cool Green Science* (blog), Nature Conservancy, June 8, 2015. http://blog.nature.org.
33. Quoted in Liz Minchin, "Air Pollution Casts a Cloud over Coral Reef Growth," The Conversation, April 8, 2013. www.theconversation.com.
34. Charles Sheppard, *Coral Reefs: A Very Short Introduction*. Oxford: Oxford University Press, 2014, p. 84.

Chapter Five: Searching for Solutions

35. Earle, "Can We Stop Killing Our Oceans Now, Please?"
36. Quoted in Cave and Gillis, "Building a Better Coral Reef."
37. Ken Nedimyer, "Bringing Life Back to Coral Reefs," CNN, March 1, 2012. www.cnn.com.
38. Nedimyer, "Bringing Life Back to Coral Reefs."
39. Quoted in Kevin Gaines and Mary Frances Emmons, "Coral Restoration Foundation Plants 'Trees' Underwater," *Scuba Diving*, August 29, 2012. www.scubadiving.com.

40. US Environmental Protection Agency, "Summary of the Clean Water Act," March 11, 2019. www.epa.gov.
41. Quoted in Bethania Palma and Alex Kasprak, "Is Trump Rolling Back Clean Water Regulations?," Snopes, February 3, 2020. www.snopes.com.
42. International Maritime Organization, "International Convention for the Prevention of Pollution from Ships (MARPOL)," 2020. www.imo.org.
43. Quoted in Anastasia Toufexis, "The Dirty Seas," *Time*, June 24, 2001. www.time.com.

1. Limit or eliminate the use of plastic such as straws, bottles, disposable cups, and other single-use items.

2. Do not dispose of household chemicals by pouring them down the sink or flushing them.

3. Be careful that gasoline does not spill when fueling lawn mowers or snowblowers.

4. If vacationing in an area that has coral reefs, take care not to touch or otherwise harm them.

5. Do not buy jewelry or other items made from coral or other marine species, such as sea turtles.

6. Keep the ocean free from debris by helping clean up beaches and other coastal areas.

7. Learn which fish are raised sustainably and avoid consuming unsustainable species (including some salmon, bluefin tuna, and wild halibut).

8. Reduce your carbon footprint by, for example, walking instead of driving or by lowering the thermostat.

9. Use fewer chemicals, such as fertilizer and pesticides, in the garden to prevent toxic runoff.

10. Become educated about combating marine pollution by joining an ocean conservation organization.

4ocean—https://4ocean.com

This for-profit company's mission is to clean the oceans and coastlines while working to stop the inflow of plastic by changing consumption habits. Proceeds from 4ocean merchandise (such as bracelets made from recycled plastic and glass) provide funds for conducting ocean and beach cleanup operations, educating the public about the importance of clean oceans, and researching new methods of pollution mitigation.

Great Barrier Reef Marine Park Authority—www.gbrmpa.gov.au

This official website for the Great Barrier Reef is loaded with information on the reef; its history, animals, and ecosystems; and its health and threats to the corals. The website hosts links to images, publications, and other resources for students.

Marine Debris Program, National Oceanic and Atmospheric Administration (NOAA)—www.marinedebris.noaa.gov

This government website describes NOAA's efforts to combat marine pollution. It contains information on the world's garbage patches and NOAA's emergency response plans for cleanup of oil and other ocean pollution. It also provides activities, fact sheets, infographics, downloadable posters, and other student resources.

National Ocean Service, National Oceanic and Atmospheric Administration (NOAA)—www.oceanservice.noaa.gov

NOAA's National Ocean Service is the nation's ocean and coastal agency. Its website provides numerous facts about ocean life and ecosystems, economic impact, and more. Its "Just for Kids" page provides activities, games, videos, and plenty of interesting ocean facts.

Ocean Conservancy—www.oceanconservancy.org

Ocean Conservancy works toward science-based solutions for a healthy ocean and the wildlife and communities that depend on it. Its website offers tips on protecting the oceans and marine life and presents information on the ocean and climate, sustainable fishing, and ocean pollution.

For Further Research

Books

Michiel Roscam Abbing, *Plastic Soup: An Atlas of Ocean Pollution*. Washington, DC: Island, 2019.

Valerie Bodden, *The Deepwater Horizon Oil Spill*. Mankato, MN: Creative Education, 2019.

David de Rothschild, *Plastiki: Across the Pacific on Plastic: An Adventure to Save Our Oceans*. San Francisco: Chronicle, 2011.

Marcus Eriksen, *Junk Raft: An Ocean Voyage and a Rising Tide*. Boston: Beacon, 2017.

Carol Hand, *Coral Reef Collapse*. Minneapolis: ABDO, 2018.

Don Nardo, *Planet Under Siege: Climate Change*. San Diego, CA: ReferencePoint, 2020.

Danielle Smith-Llera, *Trash Vortex: How Plastic Pollution Is Choking the World's Oceans*. North Mankato, MN: Compass Point, 2018.

Judith S. Weis, *Marine Pollution: What Everyone Needs to Know*. New York: Oxford University Press, 2015.

Internet Sources

India Block, "Four Technologies Tackling the Problem of Plastic Pollution in Rivers," Dezeen, November 29, 2019. www.dezeen .com.

Drew Brucker, "Ocean Pollution Facts, Stats, and Solutions," Rubicon, October 9, 2017. www.rubiconglobal.com.

Sybil Bullock, "Meet the Terrific Teens with Pollution Solutions!," Greenpeace. www.greenpeace.org.

James Ellsmoor, "Cruise Ship Pollution Is Causing Serious Health

and Environmental Problems," *Forbes*, April 26, 2019. www
.forbes.com.

NOAA Marine Debris Program, "Garbage Patches," March 7, 2020.
www.marinedebris.noaa.gov.

Ocean Portal Team, "Gulf Oil Spill," Smithsonian Ocean, April
2018. www.ocean.si.edu.

Wallace White, "Her Deepness," *New Yorker*, July 28, 2014. www
.newyorker.com.

Index

Picture Credits

Cover: ugurhan/iStock

8: kajornyot wildlife photography/Shutterstock.com
13: Maury Aaseng
17: Paulo de Oliveira/NHPA/Avalon.red/Newscom
19: Rich Carey/Shutterstock.com
22: Xinhua Xinhua News Agency/Newscom
26: iStock
28: rbrown10/Shutterstock.com
32: Vintagepix/Shutterstock.com
36: Christopher Meder/Shutterstock.com
39: EyeMark/Deositphotos
43: Vlad61/Shutterstock.com
46: iStock
49: stockphoto-graf/Shutterstock.com
55: COVER Images/ZUMA Press/Newscom
58: John Nicholls Photography/Shutterstock.com
60: Carey Wagner/ZUMA Press/Newscom

About the Author

Craig E. Blohm has written numerous books and magazine articles for young readers. He and his wife, Desiree, reside in Tinley Park, Illinois.